PHYSICAL SCIENCE

How Magnets Work

MICHÈLE DUFRESNE

TABLE OF CONTENTS

Magnets 2
What Is Magnetism? 4
What Are Magnetic Poles? 8
Is the Earth a Magnet? 12
How Are Magnets Used? 16
Glossary/Index 20

PIONEER VALLEY EDUCATIONAL PRESS, INC

MAGNETS

Have you ever played with magnets? Were you surprised when they snapped together and stuck like glue? Did you feel the **force** when you held them close?

Magnets are all around us every day.

WHAT IS MAGNETISM?

Magnetism is a form of energy. Magnetism can make metal objects spin, jump, or stick to other metals.

All magnets have the ability to pull things toward themselves but you cannot see this force. This **invisible** force is called magnetism. When an object is magnetic, it **attracts** some metals by pulling them toward the magnet.

MORE TO EXPLORE

Magnets do not attract all metals. For example, **gold, silver, and copper are not attracted to magnets**. Hold a penny to a magnet and see if it sticks. You will find that pennies are not attracted to magnets because they are made of copper.

5

Magnets can move some objects without touching them. If you move a metal paper clip slowly toward a magnet, the paper clip will jump to the magnet.

As you bring a metal object closer to a magnet, the attraction, or force, between them will grow stronger.

MORE TO EXPLORE

Magnets have different strengths. Stronger magnets can move **heavier objects.**

WHAT ARE MAGNETIC POLES?

A magnet has two ends called poles.
The poles may look the same,
but they behave differently.

When the poles of two magnets touch,
they will either stick together
or push away from each other.
You may feel a pulling force
as the two poles stick together.
Or you may feel a pushing force
as the two poles push away from each other.

The poles of these magnets are located at their ends. The areas around the poles are called magnetic fields. The tiny pieces of metal in this picture show the shape of a magnetic field.

One pole is called the north pole, and the other is called the south pole. The north pole of a magnet will attract the south pole of other magnets. When you place the same poles of two magnets near each other (north to north or south to south), they will **repel**, or push apart from, each other.

IS THE EARTH A MAGNET?

Did you know that the earth has two magnetic poles? One is close to the North Pole and one is close to the South Pole.

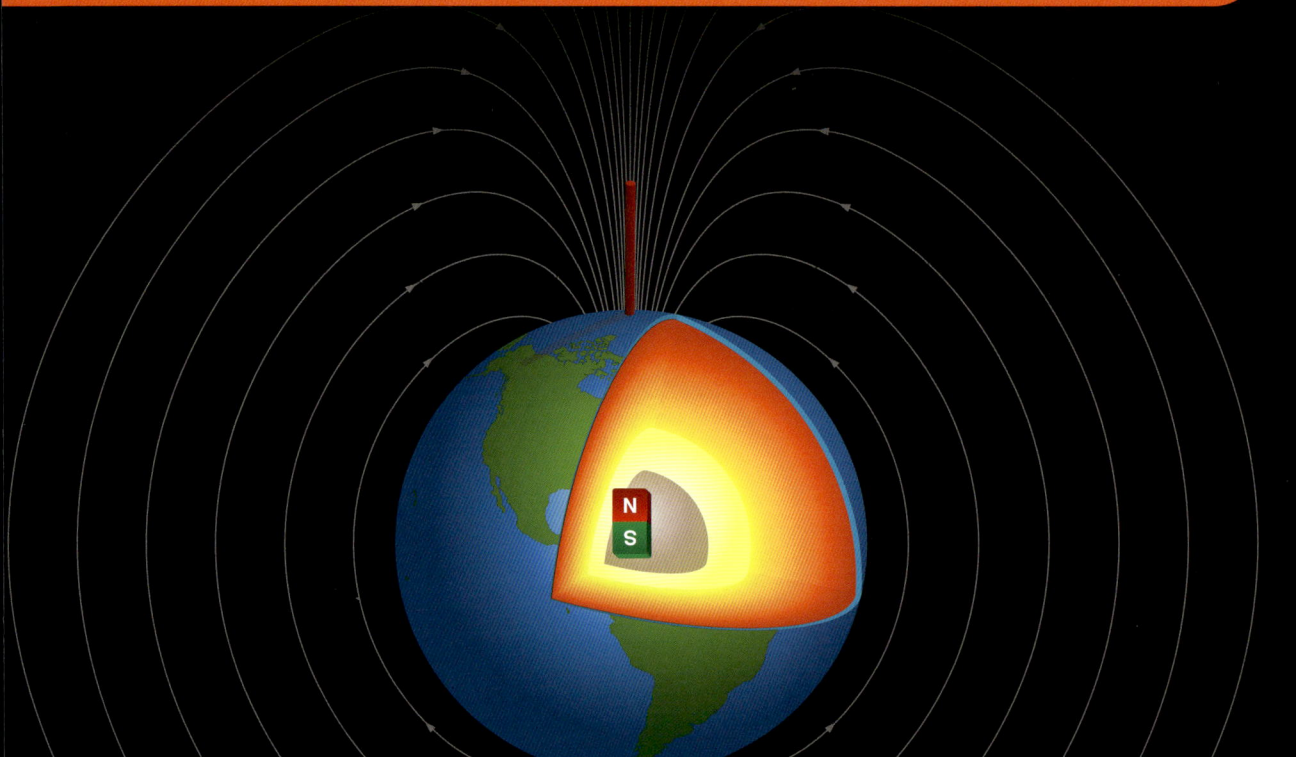

The earth's magnetic force is not very strong. The small magnets you have in your home or at school have more magnetic force.

A **compass** is a tool that uses the earth's magnetic field to show you which direction you are facing. Inside the compass is a magnetic needle that always points to the north. If you are out on a walk, a compass can help you find your way.

HOW ARE MAGNETS USED?

Magnets can be useful in many ways.

Many refrigerator doors are made of steel. You may be able to stick a note or picture to your refrigerator with a magnet.

Recycling centers use magnets to sort trash! Large magnets can separate metal objects from glass and paper.

MORE TO EXPLORE

Magnets will not pick up soda cans. Soda cans are made of **aluminum**, and magnets are not attracted to aluminum.

Some shower curtains have small magnets tucked inside their bottom stitches. The magnets keep the shower curtains inside the tub so you do not end up with water spraying on the floor.

One of the most powerful magnets is used in hospitals. Doctors use MRI (Magnetic Resonance Imaging) machines to look inside the human body.

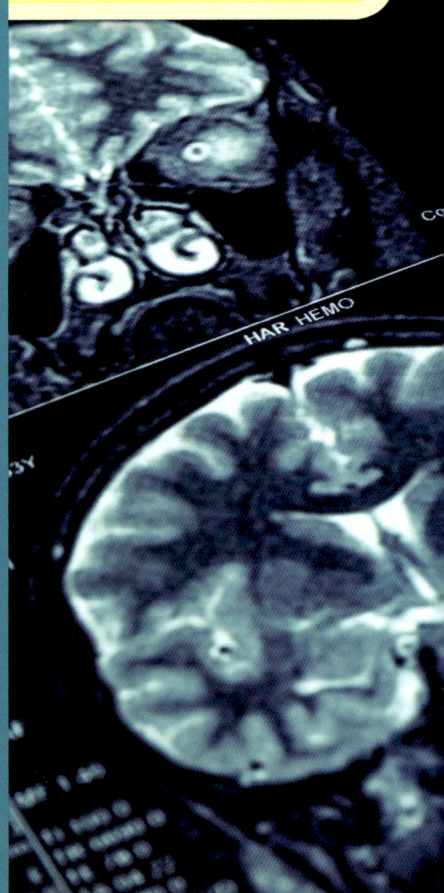

Pushing, pulling, holding, turning—there are magnets everywhere you look. No matter where you are, you live in a magnetic world!

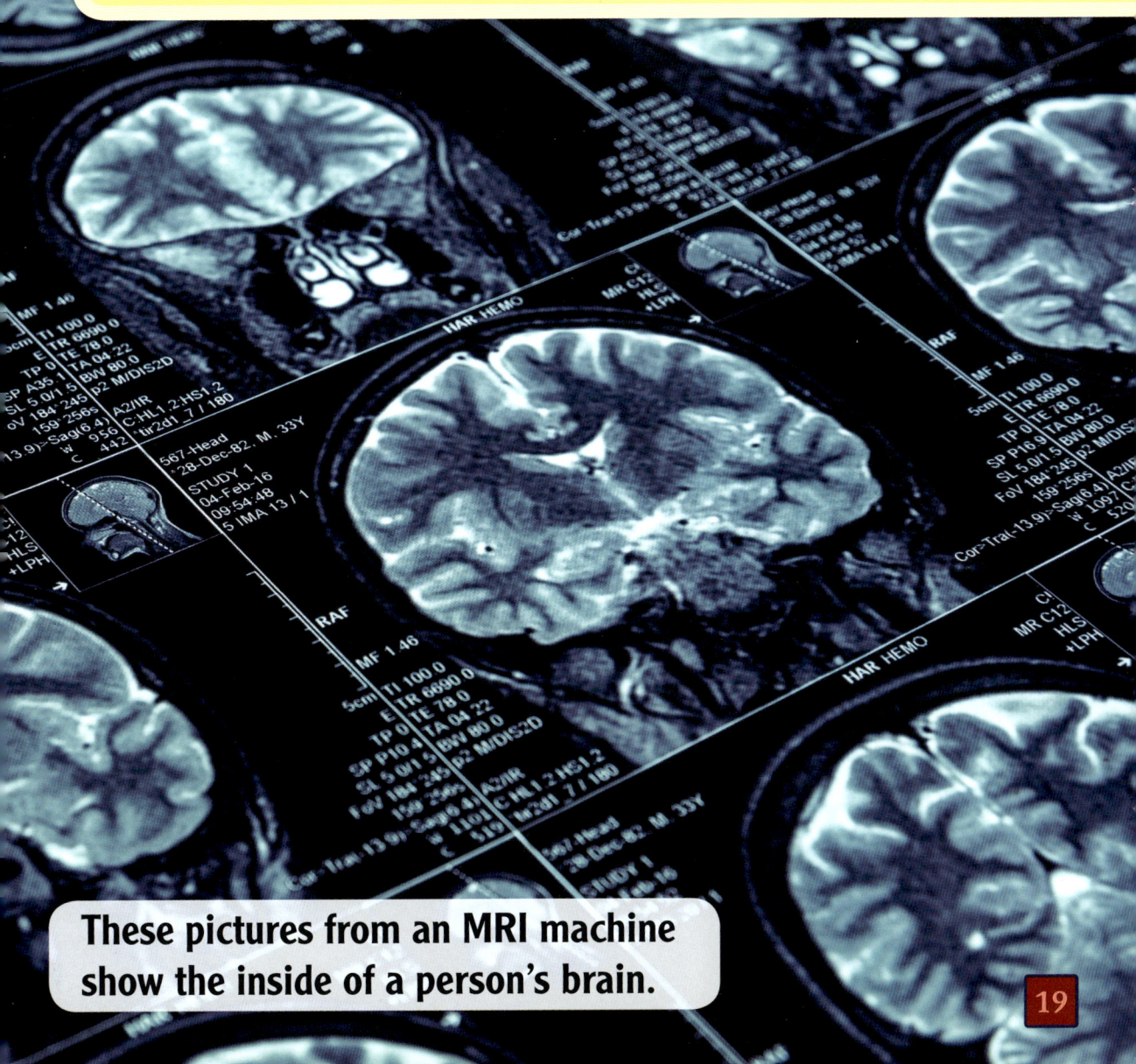

These pictures from an MRI machine show the inside of a person's brain.

COMPASS

WHAT YOU NEED:

- **sewing needle**
- **magnet**
- **wax paper**
- **scissors**
- **water**
- **bowl**
- **pen**

STEP 1 Magnetize the needle by rubbing the magnet on it from end to end.

STEP 2 Cut out a 1-inch circle of wax paper with the scissors. Then weave the needle into the wax paper and mark the pointy end of the needle with an *N*.

STEP 3 Fill the bowl with water and float the compass in the water.

Your needle should spin until it stops and points north. Check it against an actual compass.

HOW TO MAKE YOUR OWN

GLOSSARY

aluminum
a strong, light metal often used to make soda cans

attracts
causing an object to move toward something

compass
a device used to find out the direction that you are facing

force
a natural power that can change the way something moves

invisible
cannot be seen

magnetism
a force that attracts certain metals

repel
to force (something) to move away or apart

INDEX

aluminum 17
attraction 6
attracts 4, 5, 10, 17
compass 14
copper 5
energy 4
force 2, 4, 8, 13
gold 5
invisible 4
magnetic field 9
magnetism 4, 12
MRI 18
needle 14
north pole 10
North Pole 12
paper clip 6
penny 5
poles 8-10, 12
radio waves 18
refrigerator 16
shower curtain 18
silver 5
soda can 17
south pole 10
South Pole 12